Chicken Alphabet

Elizabeth Green

Book Wyrm
PUBLISHING

Published by Book Wyrm Publishing
www.bookwyrmpublishing.com

ISBN: 979-8-9928577-0-2

A 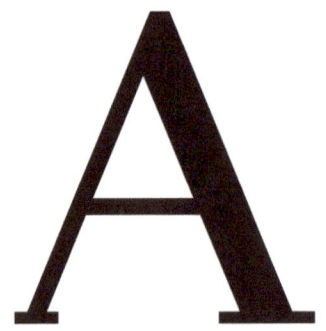 Austrolorp

A popular black-feathered chicken breed.

B

Broody

A broody hen wants to hatch eggs and stay in her nest even if there are no eggs!

C Comb

The red, fleshy part on top of a chicken's head.

D Dust Bath

Fluffing dirt into their feathers helps chickens get rid of parasites.

E

Eggs

Chickens lay eggs in many colors!

F

Feathers

Chickens feathers come in many colors and patterns, such as barred, pencil, and lace.

G

Gizzard

A special organ in the chickens' throat that helps them digest food.

H

Hen

An adult female chicken.

I Incubator

A machine that keeps eggs warm until they hatch if there is no hen to sit on them.

J

Jungle Fowl

The wild ancestor of domestic chickens.

K

Kerfuffle

Chickens sometimes squabble over food and space!

L Laying Hen

A hen that regularly lays eggs. Most hens will lay for 3-6 years.

M Molting

When chickens lose old feathers and grow new ones.

N Nesting Box

Boxes inside of a chicken coop, where hens can lay their eggs.

O Orpington

A fluffy and friendly chicken breed from England.

P Pecking Order

The social ranking of chickens in a flock.

Quirky

Chickens have unique and funny personalities!

R

Rooster

An adult male chicken. Roosters protect the flock from predators.

S Scratch

Chickens scratch the ground to uncover bugs and seeds.

T Tailfeathers

Some chickens have big fancy tails!

U Uropygial Gland

An oil gland located at the base of the tail, which chickens use to preen. Preening keeps their feathers waterproof.

V

Vent

Where eggs come out!

W Wattle

The red, fleshy parts hanging under a chicken's beak.

X  Xanthophyl

A pigment in plants that makes egg yolks yellow.

Y

Yolk

The yellow part of an egg that is full of nutrients.

Z

Zzzzzzz...

Chickens like to sleep perched on roosts!

What other chicken words do you know?

To find more fun books like this one, visit:

www.bookwyrmpublishing.com

@bookwyrmpublishing

Book Wyrm
PUBLISHING